AI AWARENESS SERIES

AI in Customer Experience

Darian Batra

Contents

Introduction

Customer expectations are evolving faster than ever—and artificial intelligence is playing a central role in that transformation. From personalized recommendations to instant support via virtual assistants, AI technologies are redefining how businesses connect with their customers across every touchpoint.

AI in Customer Experience is part of the AI Awareness Series, designed to help professionals understand and apply artificial intelligence in real-world, industry-specific contexts. This volume focuses on one of the most dynamic and competitive areas of business today: customer experience.

Great customer experiences are no longer optional—they're a key differentiator. In an era of on-demand services, digital-first interactions, and increasingly empowered consumers, organizations must find new ways to anticipate needs, personalize communication, and build lasting relationships. AI offers powerful tools to support these goals, but only if used thoughtfully, strategically, and responsibly.

In this book, we explore how AI is enabling organizations to:

Analyze customer behavior and sentiment in real time

Deliver personalized journeys across channels and devices

Enhance service with chatbots and conversational AI

Predict churn, tailor loyalty strategies, and improve retention

Create scalable, responsive, and adaptive CX operations

Ensure ethical data use and maintain customer trust

You don't need to be a data scientist to understand or benefit from AI in CX. This book is written for professionals across marketing, customer service, digital strategy, UX, and leadership—anyone seeking to improve customer relationships with intelligent tools and thoughtful design.

Throughout the chapters, we blend practical use cases with strategic insight, guiding you through the opportunities, challenges, and best practices shaping AI-enhanced customer experience today.

Let's begin.

Chapter 1: Understanding AI in CX

Customer experience covers every single interaction a customer has with a business—whether it's online, in-store, or through customer support channels. The goal of customer experience is to maximize satisfaction by consistently meeting or exceeding expectations. But it's not just about solving problems—it's about building a meaningful emotional connection that encourages loyalty and long-term relationships.

Artificial intelligence is changing the very definition of customer experience. Traditionally, companies used a reactive approach—waiting for issues to arise and addressing them manually. But with AI, customer experience becomes proactive. Businesses can now anticipate customer needs, respond in real-time, and streamline workflows using data-driven insights.

Traditional customer relationship management focused on maintaining records and handling customer interactions in a structured way. While helpful, this was often reactive, responding only after a customer made contact.

Now, with AI, we've moved toward real-time, predictive engagement. AI can instantly analyze customer data to generate insights on the spot. It can predict customer behaviors and personalize experiences in the moment, enhancing satisfaction and making interactions more meaningful.

Predictive technologies offer some powerful benefits. They can significantly boost customer satisfaction by anticipating needs and tailoring experiences. They improve operational efficiency by speeding up processes and reducing delays. However, these systems rely heavily on high-quality data—without accurate data, predictions suffer. And of course, businesses must manage privacy concerns carefully to protect trust and comply with regulations.

For AI to be effective in customer experience, businesses must prioritize data collection and integration. Gathering a wide range of data helps create a comprehensive picture of customer preferences and behaviors. Accuracy is key—flawed data means flawed insights. And integrating data from different systems gives AI a unified view, which improves its ability to make smart decisions.

Big data and analytics let businesses recognize patterns in customer behavior. This helps improve decision-making and allows companies to tailor their services and marketing to customer preferences. With the right insights, businesses can even anticipate customer needs before they're expressed, delivering proactive and highly personalized experiences.

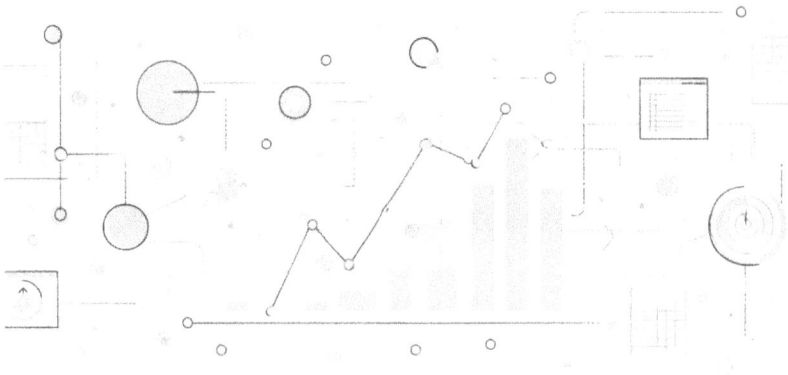

Ethical considerations are crucial when using AI in customer experience. Transparency—being clear about how customer data is used—helps build trust. Obtaining customer consent is not just a legal requirement but an ethical one. And businesses must stay compliant with privacy laws and regulations to protect data and use AI responsibly.

It's important to understand the difference between automation and AI. Automation handles repetitive tasks based on fixed rules—it doesn't learn or adapt. AI, on the other hand, learns from data, adapts to new situations, and delivers intelligent responses that go beyond pre-set programming.

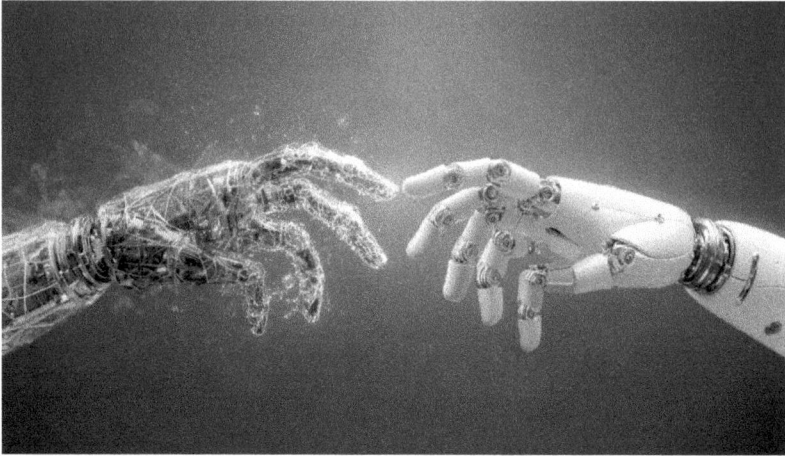

In practice, both automation and AI have valuable roles. Automation can handle routine customer service tasks quickly and efficiently. AI, especially in chatbots, provides intelligent 24/7 support that can understand and respond to complex queries. It also powers predictive support and personalized recommendations, enhancing the customer experience in ways automation alone can't achieve.

Choosing the right mix of AI and automation depends on your customer experience goals. Automation is perfect for routine, repetitive tasks, while AI is better for personalized, dynamic interactions. Finding the right balance helps optimize customer satisfaction and operational efficiency.

To wrap up, embracing AI in customer experience isn't just about adopting new technology—it's about transforming how businesses connect with their customers. By leveraging AI for proactive engagement, personalized service, and intelligent insights, companies can deliver experiences that build loyalty, drive satisfaction, and create long-term value.

Chapter 2: AI Tools and Platforms for CX

Machine learning models are at the heart of modern customer experience strategies. They power systems that can analyze data, recognize patterns, and make predictions, all of which help businesses personalize and improve customer interactions.

Let's start with personalization and recommendation systems. These models analyze customer behavior and preferences to suggest products, services, or content that align with individual tastes. By delivering highly relevant recommendations, companies can enhance customer satisfaction and boost sales.

Predictive analytics takes things a step further by forecasting customer behavior based on historical data. This helps businesses anticipate needs, plan targeted marketing efforts, and make smarter decisions about customer engagement strategies.

Churn prediction models are vital for customer retention. These models identify customers who are likely to leave, allowing companies to act before it's too late. By using these insights, businesses can launch targeted retention campaigns to address customer concerns and encourage loyalty. Ultimately, effective churn management not only improves customer loyalty but also protects long-term revenue.

Natural language processing—or NLP—helps computers understand human language. This technology powers various customer service tools, making interactions more natural and efficient.

Chatbots and virtual assistants are some of the most visible NLP applications. They offer instant, round-the-clock support, quickly handling common questions and freeing human agents to deal with complex issues. By improving response times and reducing workload, these tools enhance overall customer satisfaction.

Sentiment analysis is another important NLP application. It evaluates customer feedback to gauge satisfaction levels and detect potential problems early. By systematically processing reviews and comments, businesses can fine-tune their products and services and address customer concerns proactively.

Generative AI is changing the game in automated content creation. For example, it can generate personalized emails at scale, keeping customer communication both efficient and personal. It also helps with social media management by crafting consistent, engaging posts. And by automating responses to customer inquiries, it streamlines communication while maintaining a high level of service.

Computer vision is another exciting field with direct applications in customer experience, especially in retail and product discovery.

Visual search makes product discovery easier and more intuitive. Instead of typing keywords, customers can upload images to find matching products. This leads to better search results and improved conversion rates as customers find exactly what they're looking for more quickly.

In-store analytics powered by computer vision track customer behavior within physical stores. By monitoring movements and interactions, businesses can optimize store layouts and tailor marketing strategies based on actual customer behavior. This data-driven approach enhances both the shopping experience and sales performance.

Automated checkout systems, also using AI, streamline the payment process, reducing queues and improving customer satisfaction. At the same time, computer vision helps with security by detecting theft and misplaced items, making stores safer and more efficient.

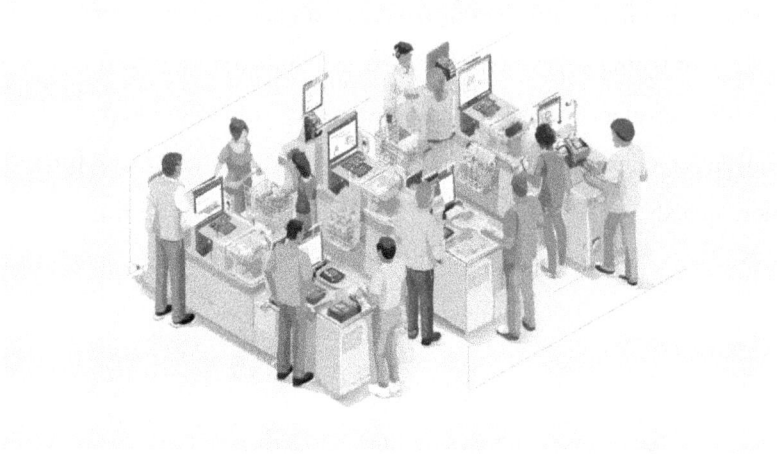

Data infrastructure plays a crucial role in supporting AI-driven customer experience strategies. Data warehouses and lakes help businesses store, manage, and analyze the vast amounts of information needed for AI systems to function effectively.

Data warehouses are designed to consolidate structured customer data from different sources. They support fast querying and reporting, enabling timely insights for customer experience analysis. With accurate, unified data, businesses can make informed strategic decisions to enhance customer interactions.

Data lakes complement warehouses by handling unstructured data—like social media posts or customer reviews. This allows businesses to perform deeper analysis on customer sentiment and behavior, tapping into a wider range of insights than structured data alone can provide.

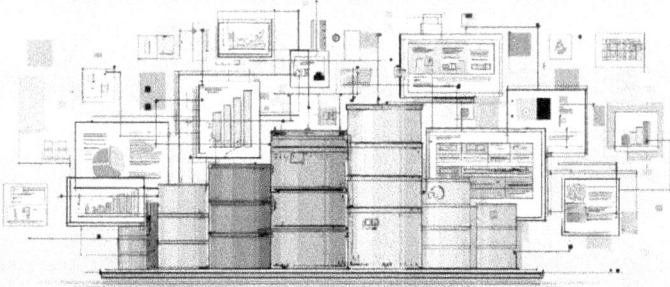

Integration of data sources is key for a holistic view of the customer. By combining data and enabling real-time processing, businesses can respond immediately to customer actions and market changes. This capability allows for dynamic personalization, ensuring each customer experience is timely, relevant, and engaging.

To wrap up, we've explored how AI technologies like machine learning, natural language processing, computer vision, and data infrastructure work together to transform customer experience. These tools empower businesses to deliver more personalized, efficient, and engaging interactions—helping them build stronger customer relationships and stay competitive in an evolving market.

Chapter 3: Real-Time Personalization

Conversational AI technologies are built on several core components. First is natural language processing, or NLP, which allows chatbots and voicebots to accurately interpret and understand user input across different languages. Then we have machine learning, which continuously improves the bot's responses by learning from past interactions. And finally, speech recognition technology enables voicebots to convert spoken words into text, making real-time voice communication possible.

Let's look at the key differences between chatbots and voicebots. Chatbots primarily use text to interact with users, often on messaging platforms. Voicebots, on the other hand, depend on speech recognition and synthesis for spoken conversations, but they face unique challenges—like dealing with unclear speech or background noise. One advantage of chatbots is their strength in multitasking and supporting asynchronous communication, which makes them highly efficient for various tasks.

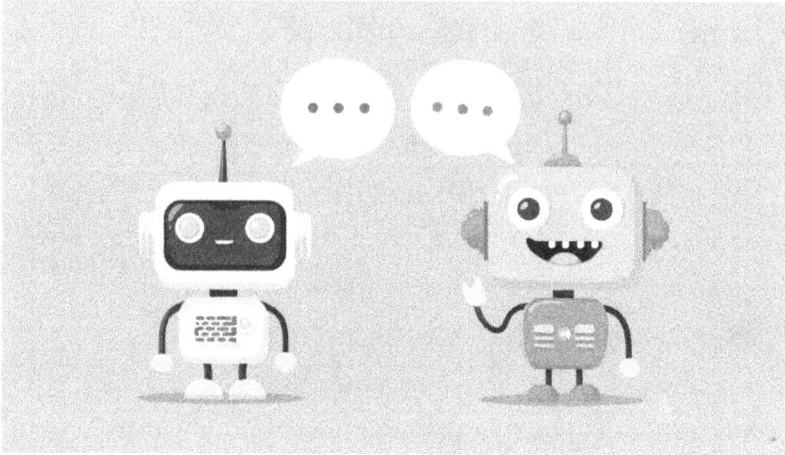

Good conversational UX design is critical for creating effective agents. It starts with clarity—making sure messages are simple and easy to understand. Conversations should also be concise to hold user interest. Being aware of the user's context allows the bot to respond appropriately and naturally. Finally, guiding users smoothly through tasks with seamless interactions helps ensure a positive experience.

Training conversational agents relies heavily on machine learning techniques. These models learn and improve over time, especially when they incorporate feedback from real users. This cycle of continuous improvement ensures that the agents become more accurate, effective, and user-friendly with every interaction.

For conversational AI to truly understand users, it must recognize context. Natural language processing techniques help track dialogue history, allowing the bot to remember what's been said and maintain context throughout a conversation. NLP also aids in disambiguating user intent—figuring out exactly what the user means based on language cues. This ability supports coherent, multi-turn conversations that feel natural and connected.

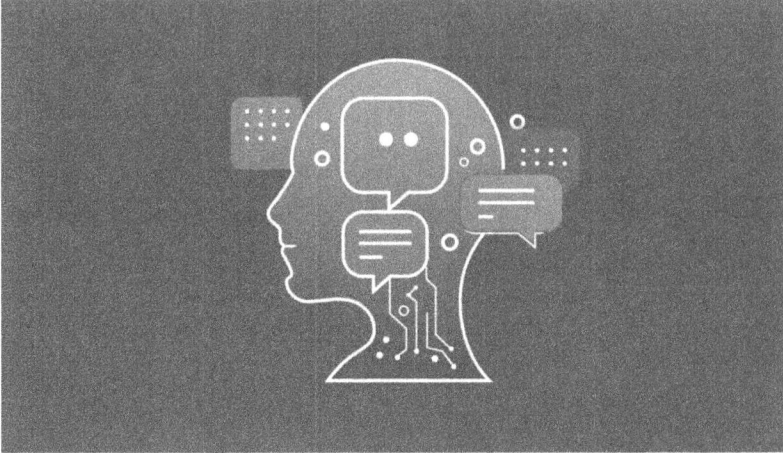

Sentiment analysis and emotion detection take contextual understanding even further. By recognizing the user's mood and emotional state, conversational agents can adapt their responses to be more empathetic and relevant. This approach leads to better personalization, making interactions more meaningful and improving the overall user experience.

Multilingual conversational AI presents its own set of challenges. Handling linguistic nuances means understanding subtle differences in meaning across languages. Cultural differences also come into play— AI systems must be sensitive to these to respond appropriately. And, of course, the AI needs to maintain high intent accuracy, ensuring that it correctly understands users no matter the language they're speaking.

An effective omnichannel strategy ensures that users have a consistent experience whether they're interacting through a website, a mobile app, or by voice. This means delivering the same quality of information, maintaining a uniform conversational style, and ensuring the user journey feels seamless across all platforms.

To engage a global audience effectively, best practices include localization—adapting language and interfaces to fit local preferences—and cultural adaptation, which respects local customs and expectations. It's also important to build scalable architectures that can handle increased demand and support growing user bases worldwide.

Large Language Models, or LLMs, have transformed conversational AI. They use vast datasets to learn complex language patterns and leverage deep learning techniques to understand nuanced inputs. When integrated into AI systems, LLMs can produce responses that are natural, coherent, and contextually relevant, making conversations feel more human-like.

LLMs significantly enhance conversational quality. They help maintain dialogue coherence, making conversations flow logically and naturally. They also bring creativity to responses—introducing variety and engaging content. Plus, their adaptability to different topics and user inputs ensures flexible, dynamic interactions that keep users engaged.

However, integrating LLMs comes with important ethical considerations. LLMs may reflect biases from their training data, which can affect fairness and inclusivity. There's also a risk of misinformation, making validation critical. Privacy concerns arise from data handling, so strict policies must be in place. Ultimately, responsible design and continuous monitoring are essential to safely and ethically deploy LLM-based solutions.

To wrap up, we've covered the essential aspects of intelligent chatbots and voicebots—from their design and contextual understanding to multilingual communication and LLM integration. As these technologies advance, keeping best practices and ethical considerations in mind will be key to creating effective and responsible AI-driven interactions.

Chapter 4: Conversational AI

AI is fundamentally reshaping customer support by introducing intelligent automation. Automated responses handle common queries, speeding up resolution times. AI's data analysis capabilities help businesses gain insights and improve personalization. Agent assistance tools provide real-time suggestions and relevant information during interactions. And by streamlining workflows, AI delivers faster and more personalized support, making both customers and agents more efficient.

AI brings many benefits to customer support. It reduces costs, improves efficiency, and provides continuous 24/7 service. But adopting AI also comes with challenges—like complex integrations, technical hurdles, and ensuring strict data privacy. Another key factor is human-AI collaboration. For AI to deliver its full potential, it must work seamlessly alongside human agents, enhancing rather than replacing their capabilities.

AI uses natural language processing to understand customer inquiries, categorize them accurately, and prioritize them based on urgency. This ensures each inquiry is directed to the right department or agent, reducing response times and streamlining the service process. With smarter routing, businesses can resolve issues more quickly and efficiently.

Automated assignment plays a key role in improving both response times and customer satisfaction. By minimizing wait times and ensuring inquiries reach the right person on the first try, AI helps increase first-contact resolution rates. This leads to happier customers who feel their issues are handled swiftly and effectively, boosting loyalty in the long run.

Real-time information retrieval allows agents to access the right data exactly when they need it. AI analyzes ongoing conversations and provides relevant knowledge base articles and solution recommendations. This not only saves time but also ensures that agents are giving customers the best possible support without having to dig for information manually.

With contextual knowledge suggestions, AI tailors article recommendations based on the customer's inquiry. This helps both customers and agents find accurate answers faster. By cutting down the time spent on research, AI improves efficiency and helps deliver quicker resolutions during support interactions.

AI also plays a big role in reducing training time and minimizing errors. Through guided assistance and real-time suggestions, new agents can learn on the job more effectively. AI helps reduce onboarding times and provides the right information when it matters, which means fewer mistakes and better service, even from less experienced agents.

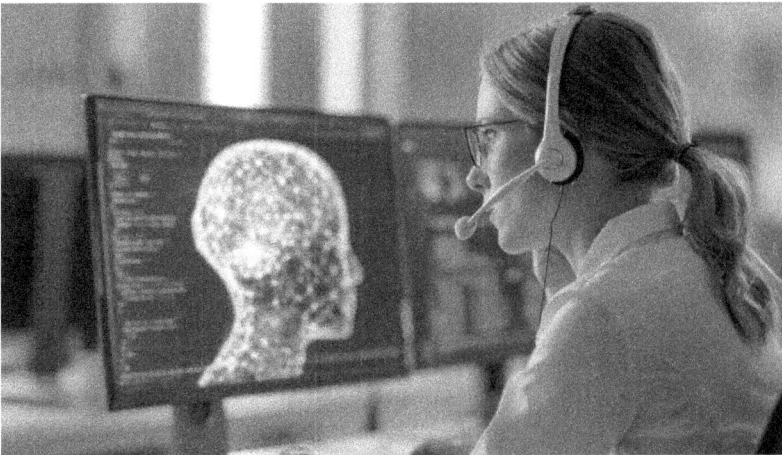

AI can detect customer sentiment by analyzing their messages through natural language processing. This helps agents pick up on emotions like frustration, happiness, or anger. By understanding customer tone, agents can adjust their responses, leading to better engagement and more effective communication.

Real-time AI coaching supports agents during live interactions. AI offers instant tips on how to communicate more effectively and even suggests better language choices. These live suggestions help agents maintain clear, positive communication, which improves the overall customer experience.

Empathy is key in customer support. By making agents more aware of customer emotions, AI helps them respond with understanding and care. This not only improves communication but also builds trust with customers, strengthening relationships and encouraging loyalty.

AI models help identify high-risk or complex cases by recognizing patterns in customer data. This allows potential escalations to be flagged early on. By spotting these situations proactively, AI ensures that the right attention is given to critical cases before they escalate into bigger problems.

AI doesn't just detect issues—it can also trigger automated escalation workflows. When certain risk factors are detected, AI initiates predefined protocols, bringing in senior agents or managers at the right moment. This timely involvement helps resolve issues faster and more effectively.

By managing escalations proactively, AI reduces time-to-resolution and improves customer outcomes. Quicker involvement of the right people means faster problem-solving, which leads directly to higher customer satisfaction. Customers appreciate quick, effective service, and AI helps make that happen consistently.

To wrap up, AI-powered customer support is revolutionizing how businesses interact with customers. From automating routine tasks to providing real-time coaching, AI enhances both the efficiency of support teams and the quality of customer service. While there are challenges, the benefits of adopting AI—like faster resolutions, better customer experiences, and increased operational efficiency—make it a game-changer in today's service landscape.

Chapter 5: Predictive Analytics for CX

Real-time personalization engines are systems that dynamically adjust content based on user data and behavior. The key is that this happens in real time—so the experience feels personal and responsive. This can lead to higher engagement as users see content that resonates with them right away.

From a business perspective, real-time personalization leads to higher conversions and stronger customer loyalty. Consumers benefit by seeing content that actually interests them, enjoying faster access to relevant information, and receiving offers that match their needs. It's a win-win situation when done right.

Several technologies make real-time personalization possible. Real-time data processing allows systems to react instantly. In-memory databases keep data easily accessible for quick responses. AI and machine learning help predict user preferences, while fast APIs ensure smooth, real-time interactions between different systems.

Behavioral profiling uses several techniques to track and analyze user behavior. Cookies track visits and preferences. Device fingerprinting uniquely identifies devices. Session tracking monitors what users do on a site in real time. And event-based analytics focuses on specific user actions, like clicks, to build detailed profiles.

By leveraging behavioral data, systems can instantly adapt content as users interact with them. This means websites and apps can update in real time to reflect a user's interests, making the experience feel more relevant and personal—and ultimately increasing engagement.

Here are some common examples of dynamic content in action. Personalized homepages adjust based on user preferences. Adaptive banners change depending on what users show interest in. Product recommendations are tailored for each user. And chatbots respond personally based on the user's specific questions and actions.

AI-driven segmentation often starts with machine learning models that identify patterns in user data. These models group users into meaningful segments based on shared behaviors or preferences. This allows businesses to target audiences more effectively with personalized experiences.

Predictive analytics helps businesses identify high-value users—those most likely to convert or engage. By analyzing both past and real-time data, these models help marketers focus their efforts, maximizing the return on investment for their campaigns.

Personalized messaging is about more than just using a customer's name. AI tailors the content of messages, selects the best time and channel to reach the user, and even adjusts offers based on individual preferences—resulting in higher engagement and better conversion rates.

There are three main types of recommendation algorithms. Collaborative filtering suggests items based on the behavior of similar users. Content-based filtering looks at item attributes to make recommendations. Hybrid approaches combine both methods to improve the relevance and accuracy of suggestions.

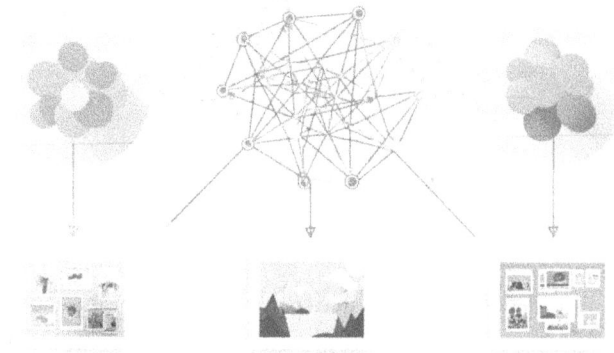

Recommendation systems are used in many industries—from e-commerce platforms suggesting products, to media services recommending content, and even in customer support for delivering helpful resources. Their flexibility makes them a powerful tool in various service areas.

However, recommendation systems face some big challenges. They need to handle massive amounts of data efficiently. They must ensure diversity and fairness in their suggestions to avoid bias. And most importantly, they must build and maintain user trust through transparency and ethical practices.

Generative AI is transforming personalization by creating unique content on demand. This allows businesses to deliver hyper personalized experiences, like product descriptions or messages crafted for individual users, making every interaction feel truly unique.

Some real-world applications include AI-generated product recommendations that adapt to user behavior, dynamic storytelling experiences where narratives change based on choices, and chatbots that use natural language understanding to provide highly personalized assistance.

With great power comes great responsibility. As we use personalization and AI, we have to prioritize data privacy and user consent. Being transparent about how algorithms work and actively preventing bias are critical steps in building responsible AI systems that users can trust.

To wrap up, real-time personalization engines are reshaping how businesses engage with customers. With technologies ranging from behavioral profiling to generative AI, these systems offer immense value—but they must be used ethically and responsibly. Thanks for watching, and I hope this session gave you clear insights into the world of real-time personalization.

Chapter 6: Proactive Support and Issue Prevention

Predictive customer behavior refers to using data analytics to forecast future customer actions and preferences. This helps businesses tailor experiences, increasing customer engagement and loyalty. By predicting behavior, companies can optimize marketing efforts, ensuring campaigns are more targeted and deliver a better return on investment. And when businesses anticipate customer needs, they naturally improve satisfaction and retention.

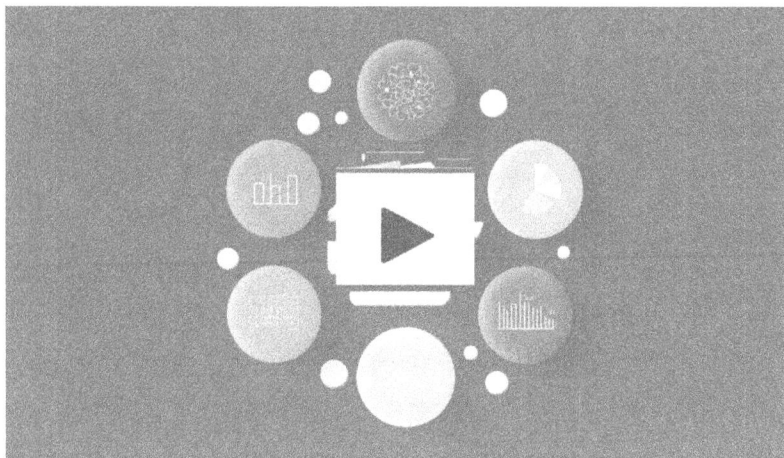

To generate these insights, companies rely on customer data — everything from purchase history to interactions and demographic information. By analyzing this data, they can identify patterns and emerging trends in customer behavior. These insights then inform strategic decisions, helping businesses engage customers more effectively and drive better outcomes.

Predictive analytics has many practical applications in business today. It powers personalized marketing by tailoring campaigns to customer preferences. It's used in inventory management to forecast demand and avoid stockouts. Companies use it for risk assessment, helping them make smarter decisions and reduce potential losses. And in customer support, predictive analytics anticipates needs, allowing businesses to offer better service.

Propensity models are built on historical customer data. They use predictive algorithms to score customers based on how likely they are to take certain actions — like making a purchase or churning. These scores help businesses intervene effectively, focusing their marketing on the right customers at the right time.

There are several key techniques for predicting customer intent. Logistic regression models are widely used for their ability to predict the likelihood of specific behaviors. Decision trees are great for segmenting customers and forecasting intent based on clear rules. And advanced machine learning algorithms can analyze complex behavioral signals, delivering highly accurate predictions that enhance targeting.

Once you have propensity scores, you can integrate them into your marketing strategies. This allows you to create highly personalized campaigns that resonate with individual customers. Propensity scoring also helps allocate marketing resources efficiently — targeting those with the highest likelihood of converting. The result? Better customer response rates and improved conversion.

Identifying at-risk customers starts with tracking engagement metrics — watching for signs of declining interaction. Analyzing transaction frequency can reveal customers who are starting to disengage. Sentiment analysis of customer feedback highlights dissatisfaction. And by applying machine learning, businesses can improve the accuracy of their churn predictions.

When it comes to predicting churn, key indicators include usage patterns, the frequency of complaints, and interactions with customer support. To gather this data, businesses draw from various sources — CRM systems, social media, and direct feedback — ensuring a comprehensive view that enhances their predictive models.

Retention strategies work best when they're informed by predictive insights. Personalized offers and targeted incentives help keep customers engaged. Loyalty programs encourage repeat business and foster stronger relationships. And by using churn risk scores, companies can tailor their communications, ultimately maximizing the lifetime value of each customer.

Next-best-action frameworks start by integrating customer context — using data to select actions based on individual behaviors and preferences. These frameworks also ensure that recommended actions align with business objectives, making them both customer-centric and commercially effective. Predictive models play a key role here, anticipating needs and dynamically suggesting the most relevant next action.

Building an effective offer engine begins with historical customer data — analyzing past behavior to understand preferences. Real-time data integration then allows offers to be adjusted on the fly, keeping them relevant. The result is tailored promotions that engage customers and drive higher conversion rates.

Customer lifetime value, or CLV, is influenced by several key factors. Purchase frequency shows how often customers buy from you. The average order value reflects how much they spend. Retention rates determine how long customers stay with your business. And acquisition costs affect your overall profitability. Together, these factors help you estimate each customer's long-term value.

There are several popular models used for predicting CLV. Some focus on historical spending and behavior, while others incorporate predictive analytics and machine learning. The goal of these models is always the same — to help businesses forecast future customer value with greater accuracy.

Once you have CLV insights, you can use them to grow your business. Start by prioritizing high-value customers — focusing your resources on the people who bring in the most revenue. Use CLV data to optimize your marketing spend, ensuring your budget is directed where it will have the greatest impact. And finally, develop personalized strategies that increase long-term profitability by strengthening customer relationships.

To wrap up, predictive customer behavior analytics offers powerful tools for understanding and influencing customer actions. By applying these advanced methods, businesses can enhance engagement, boost retention, and drive profitability. Predictive insights not only help tailor marketing and service strategies but also support smarter, data-driven decisions that build lasting customer relationships.

Chapter 7: AI in Feedback Collection

Automated analytics in surveys and reviews relies on AI algorithms to process large volumes of data quickly and accurately. This technology enables organizations to extract insights much faster than manual analysis would allow. By reducing manual effort, businesses can respond to customer feedback more efficiently, improving service delivery and customer relationships.

Automation addresses several key challenges. It provides businesses with actionable insights, allowing better decision-making and strategic growth. By tailoring services based on feedback, customer satisfaction and loyalty can be greatly improved. Automated systems also help detect customer pain points early on, leading to quicker resolutions and higher service quality.

Sentiment detection in natural language processing begins with identifying the polarity of text—whether feedback is positive, negative, or neutral. Lexicon-based methods use sentiment dictionaries for this classification, while machine learning models learn from labeled data to refine their accuracy. Deep learning models go further by understanding sentiment intensity and context, providing even deeper insights.

When analyzing voice data, it's important to extract key features like tone, pitch, and speech rate to interpret emotional cues effectively. By combining acoustic analysis with speech recognition, organizations can enhance the accuracy of sentiment detection in spoken feedback, leading to more reliable insights.

In real-world feedback systems, sentiment analysis is used across industries to understand customer attitudes, guide product improvements, and tailor marketing efforts. Whether in customer service, product development, or market research, these insights inform better business decisions and customer engagement strategies.

Topic modelling algorithms like Latent Dirichlet Allocation, or LDA, group related words to uncover hidden themes in large datasets. Another method, Non-negative Matrix Factorization, breaks down text data into word-topic matrices, helping organizations extract meaningful topics efficiently from massive amounts of customer feedback.

Topic modelling allows businesses to identify dominant themes within their datasets. By tracking how topics change over time, companies can spot emerging trends, shifting customer interests, and new concerns early, enabling proactive responses and strategic adjustments.

The insights gained from topic modelling support several business areas. For product development, it reveals customer needs and guides innovation. In marketing, it helps refine strategies by identifying customer interests. For customer support, it highlights common issues, enabling more effective service delivery and improved satisfaction.

Detecting emotions involves analyzing linguistic cues in text, such as words that signal joy, frustration, or anger. For spoken feedback, facial expression analysis and acoustic feature extraction—like tone, pitch, and volume—help identify emotions, making it possible to understand the true sentiment behind customer interactions.

Interpreting customer mood enables businesses to prioritize their responses, addressing urgent concerns promptly. It also allows communication to be tailored in a way that resonates with the customer's emotional state, fostering more meaningful engagement. By delivering empathetic service, companies can build stronger relationships and increase customer loyalty.

Integrating emotion analysis directly into feedback platforms enhances their value. Real-time mood assessment enables organizations to respond to customer emotions as feedback is collected. This enhances customer engagement and allows companies to detect dissatisfaction early, leading to quicker resolutions and improved satisfaction.

Automating the feedback response process creates a closed-loop system where customer feedback leads directly to action. This ensures that no feedback is ignored and that customer concerns are systematically addressed, supporting continuous improvement efforts within the organization.

Continuous improvement relies on three key activities. First, tracking feedback trends helps identify what needs to change. Second, organizations implement those changes based on actionable insights. And third, they monitor the effectiveness of those changes over time to ensure they're making the desired impact.

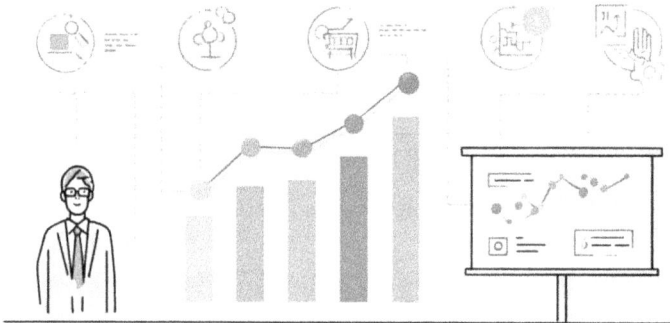

Measuring the success of closed-loop systems involves tracking customer satisfaction metrics, analyzing response times, and evaluating resolution rates. These metrics help organizations understand how well they are addressing feedback, where efficiencies can be gained, and how customer satisfaction is evolving.

To conclude, automated survey and review analysis equips businesses with powerful tools for understanding customers and improving experiences. By combining sentiment analysis, topic modelling, emotion recognition, and closed-loop feedback, organizations can make data-driven decisions, enhance service quality, and foster lasting customer relationships.

Chapter 8: Text and Speech Analytics

Voice of the Customer Analytics is all about gathering customer feedback to understand their needs and expectations. By analyzing this feedback, organizations can pinpoint areas for improvement and align their services more closely with customer needs — which ultimately leads to better customer satisfaction.

The key objectives of Voice of the Customer Analytics are to improve product quality, enhance customer experience, and drive business growth. By leveraging customer feedback, companies can make smarter, data-driven decisions that improve customer satisfaction and give them a competitive edge.

Here's a look at the typical workflow for VoC (Voice of the Customer) Analytics. First, we collect feedback from various customer touchpoints. Then, we aggregate that data into a single dataset for analysis. Sentiment analysis helps us understand how customers feel, and finally, we extract actionable insights that lead to meaningful improvements.

Customer feedback comes from many sources. Surveys provide structured insights into satisfaction and opinions. Social media gives us real-time sentiment and trends. Call centers offer detailed feedback through direct conversations, and emails and online reviews provide written records of customer experiences.

To unify feedback from all these sources, we use techniques like data cleansing to remove errors, data normalization to ensure consistency, and integration tools — like ETL processes and data lakes — to bring structured and unstructured data together for analysis.

However, aggregating data isn't without its challenges. Data silos, inconsistent formats, and large volumes of data can all pose problems. Best practices include establishing clear data standards, leveraging automation for efficiency, and maintaining strict data quality controls.

Real-time monitoring of customer feedback allows businesses to catch issues early — before they escalate. This means they can respond immediately to customer concerns, which helps prevent problems from growing and improves overall customer satisfaction.

AI plays a key role in anomaly detection. Machine learning models can spot unusual patterns in feedback data, while statistical algorithms help identify shifts in sentiment or behavior that might indicate emerging problems.

Here are some common use cases: detecting service outages before customers are seriously affected, identifying product defects early for quick action, and monitoring negative sentiment spikes to resolve customer issues proactively.

When we analyze Voice of the Customer data over time, we can identify persistent issues, see the impact of our changes, and understand how customer expectations are evolving. This long-term perspective helps guide strategy and continuous improvement.

We can extract valuable insights from VoC data using techniques like trend analysis, sentiment scoring, root cause analysis, and segmentation. These methods help us understand the bigger picture and enable more targeted, effective actions.

Real-world case studies show how organizations use VoC insights to drive strategic change. Whether it's improving products, enhancing customer loyalty, or refining services, applying these insights leads to better outcomes and stronger customer relationships.

Integrating VoC data into customer experience dashboards brings all insights into one place. This centralizes data for a full-picture view, promotes cross-functional collaboration, and enables faster, better-informed decisions across the business.

Effective VoC-enabled dashboards offer real-time data visualization, customizable metrics, drill-down capabilities for deeper insights, and automated alerts for significant changes or anomalies — all essential features for modern customer experience management.

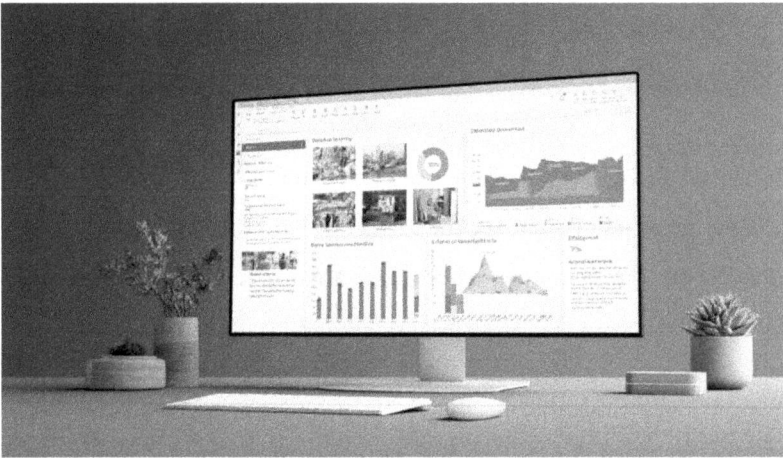

To get the most from VoC analytics, dashboards must align with business objectives. It's important to involve stakeholders in the design process, ensure the accuracy of your data, and keep dashboards updated as business needs evolve. These practices make sure your VoC efforts stay relevant and effective.

To conclude, leveraging advanced VoC analytics techniques allows businesses to deeply understand customer feedback, respond proactively, and make strategic, data-driven decisions. By integrating these insights effectively, organizations can significantly enhance both customer satisfaction and business performance.

Chapter 9: AI Readiness for CX

Journey mapping helps visualize customer experiences by showing the different interactions a customer has with a brand. It helps organizations understand these interactions and pinpoint where things go wrong. With AI integration, journey mapping goes further — using machine learning and advanced analytics to uncover deeper insights from data. This means companies can automatically optimize journeys, making the experience better for customers without relying on manual analysis.

Several technologies power AI-driven journey mapping. Machine learning is key — it spots patterns and predicts customer behaviors. Natural language processing helps understand customer feedback and conversations, even when they're unstructured. Big data analytics processes huge amounts of information to drive smart decisions. And real-time event processing ensures that journey maps are updated on the fly as customers interact with the brand.

AI makes journey mapping more accurate and scalable — allowing businesses to analyze large, complex datasets. It also helps personalize customer insights, so the experience can be tailored for each individual. But there are challenges too. Data privacy is a big concern — companies must handle customer information responsibly. AI models can also have biases, and implementing AI can be complex, requiring careful management.

AI helps detect where customers are dropping off in their journey. By analyzing user behavior, AI can spot patterns that show where people lose interest or abandon the process. This lets companies take action proactively — whether that's fixing a website glitch, changing a communication strategy, or offering help at the right moment — to keep customers engaged.

AI can also analyze behavioral data to identify pain moments — those frustrating points that cause friction for customers. By detecting these pain moments, teams can prioritize fixing the most critical issues. This makes the journey smoother and keeps customers from abandoning their interaction altogether.

AI models predict customer behavior by analyzing historical data — spotting trends that can help forecast what customers are likely to do next. When combined with real-time data, these models adjust dynamically to customer actions as they happen. This means businesses can better anticipate customer needs and guide them down preferred paths, enhancing both targeting and the overall experience.

Predictive analytics lets organizations adjust customer journeys before problems arise — ensuring a smoother experience. By making proactive changes based on AI predictions, companies can increase customer satisfaction and achieve better business outcomes, like higher sales or better engagement rates.

Let's look at real-world examples of predictive path optimization. In many cases, companies have used AI to anticipate customer needs, reducing wait times and improving service. Operationally, predictive models help businesses work more efficiently — cutting costs and using resources better. These approaches are being used across industries to solve complex problems and boost performance.

Event-driven triggers are specific actions or external events that cause a response in a customer journey. These triggers allow businesses to personalize responses — making interactions more relevant based on the customer's behavior or context at that exact moment.

AI processes event data in real time, allowing businesses to react instantly. This means customer experiences can be dynamically adjusted, keeping interactions relevant and engaging. Whether it's a personalized offer or a timely message, AI makes sure the response matches the customer's needs right when it matters.

Measuring the impact of event-based personalization is key. Metrics like customer loyalty, conversion rates, and satisfaction scores help businesses understand if their personalized experiences are making a difference. By analyzing these metrics, organizations can fine-tune their strategies for even better engagement.

Synthetic data mimics real customer data but doesn't contain actual personal information. This allows organizations to run tests and simulations without privacy concerns. Synthetic data is crucial for safely testing journey scenarios and making sure systems work as intended before being used with real customers.

AI models can be trained on synthetic data, ensuring privacy is protected during development. This data allows for varied journey scenario testing, helping teams see how different customer paths play out. It also supports accurate outcome prediction — all without risking exposure of real customer data.

Simulations using synthetic data allow organizations to test journey strategies before rolling them out. By predicting outcomes in a safe environment, companies can refine their approach and minimize risks. This data-driven improvement process means better strategies and more confident decision-making.

To wrap up, AI is transforming journey mapping by making it more insightful, proactive, and personalized. Whether it's identifying pain points, predicting behavior, or simulating new strategies, AI empowers organizations to create better customer experiences. Embracing these AI-powered techniques can lead to significant competitive advantages in customer engagement and satisfaction.

Chapter 10: Integration and Workflow Automation

Let's start with an overview of proactive and preventative approaches. Proactive service means anticipating customer needs before they arise, providing timely support and solutions. Preventative service focuses on identifying potential issues early and stopping them before they affect the customer. Both approaches are designed to reduce friction and improve responsiveness, leading to higher customer satisfaction.

Now, let's look at the key benefits these strategies bring. First, they help increase customer loyalty by fostering stronger relationships and trust. They also reduce operational costs through more efficient processes. Faster issue resolution boosts both customer satisfaction and operational performance. And, perhaps most importantly, they create a seamless experience with fewer disruptions and more personalized service.

However, implementing these strategies comes with some challenges. Data integration is critical—organizations must unify diverse data sources. Effective cross-team coordination is needed to align goals and ensure smooth execution. There's also a need for investment in advanced technologies to support these efforts. Finally, we have to balance automation with trust, making sure that while automation helps, it doesn't erode customer confidence or privacy.

Now, let's talk about AI nudging. AI nudging uses data-driven insights to understand customer behavior and preferences. It's about making subtle, timely suggestions that guide decisions without limiting choice. This leads to better engagement and satisfaction through relevant, personalized interactions.

Here are some real-world uses for timely AI-driven interventions. For example, delivering personalized offers based on customer preferences. Sending subscription renewal reminders to help with retention. Proactively alerting customers about potential issues before they escalate. And recommending support options proactively to improve customer satisfaction.

Measuring the effectiveness of AI nudging is key. We use metrics like conversion rates, customer feedback, and service incident reductions. Continuous data analysis helps refine these strategies over time. And A/B testing allows us to compare different approaches and choose the most effective ones.

Let's move on to mapping the customer lifecycle. We start with the awareness stage—when customers first learn about a product or service. Then comes consideration, as they evaluate options. Next is the purchase stage, when they make the buying decision. Finally, we have retention and advocacy, where the focus is on keeping customers engaged and encouraging them to promote the brand.

To make the most of lifecycle-triggered communications, we need to identify the right triggers. Using customer data to personalize outreach increases relevance and engagement. This leads to higher response rates and more conversion opportunities. Ultimately, personalized communication helps build long-term customer loyalty.

Predictive analytics is at the heart of automated outreach for issue prevention. By analyzing data patterns, we can detect issues before they happen and reach out proactively. This helps avoid problems before they affect the customer.

Let's look at some common techniques for automated customer outreach. Email alerts provide timely updates straight to the customer's inbox. SMS notifications deliver quick messages directly to mobile devices. Chatbots offer 24/7 automated interactions without needing human support. And in-app messaging provides personalized communication within web or mobile applications.

Automated outreach plays a key role in reducing incidents and boosting satisfaction. By addressing issues proactively, we prevent service disruptions and escalation. This timely support enhances trust, satisfaction, and ultimately, customer loyalty.

Customer experience and employee experience are deeply interconnected. When employees are satisfied and motivated, they naturally deliver better customer service. Empowered employees create positive interactions, which helps foster customer loyalty and supports business growth.

Joint journey mapping is essential to optimize both customer and employee experiences. By integrating CX and EX journeys, we identify shared touchpoints. This allows for coordinated improvements that enhance both the customer and employee experience. The result is a smoother, more consistent experience for everyone involved.

Driving value requires a holistic experience design. A comprehensive strategy takes into account the needs of both customers and employees. This leads to improved efficiency, stronger loyalty, and better business performance. Ultimately, this approach gives organizations a lasting competitive advantage.

To conclude, proactive and preventative service strategies—supported by AI nudging, lifecycle-triggered communications, and automated outreach—play a crucial role in modern customer experience. When combined with a focus on both customer and employee journeys, these approaches help businesses drive loyalty, reduce costs, and stay competitive in today's market.

Chapter 11: Human-AI Collaboration in CX

Let's start by defining what automated decision-making actually is. These are decisions made by algorithms without any human input. They're used in areas like credit scoring, where algorithms analyze financial data to assess creditworthiness, in healthcare to assist with diagnosis, and even in hiring, where algorithms help screen candidates and streamline recruitment. The key point is that these systems make independent decisions based on data.

When we think about ethics in automation, three core principles stand out. First is fairness — making sure systems treat everyone equally without bias. Then there's accountability and transparency — designers and companies must be able to explain how decisions are made and take responsibility for them. And finally, privacy and human autonomy — protecting people's data and making sure they have control over decisions that affect them. These principles help guide responsible automation.

Algorithmic personalization aims to tailor content and services to individual users. By analyzing data, algorithms customize content, improving engagement and user satisfaction. But this comes with a risk — personalization can create echo chambers, limiting the perspectives users are exposed to and potentially reinforcing their existing biases.

Bias in algorithms often comes from skewed training data — if the data isn't representative, the model will reflect those imbalances. Poor model design can also embed bias into the system, and biased feedback loops make it worse by reinforcing skewed outcomes over time. The consequences? Unfair treatment, discrimination, and a loss of trust in technology systems.

To tackle bias, we need diverse, representative training data. Bias detection tools can help flag issues before deployment. Transparency is also critical — when stakeholders can see how algorithms make decisions, it's easier to hold them accountable. And regular audits are a must to catch unfair outcomes and correct them before they cause harm.

Now, let's look at automated agent routing. This is where algorithms assign customer inquiries to service agents. The goal is to optimize efficiency and customer satisfaction, but if not handled carefully, this can result in unfair service allocation — for example, some customers consistently getting lower service levels due to algorithmic decisions.

Preventing unfair service allocation starts with monitoring. By constantly reviewing service metrics and gathering feedback, we can catch biases early. Building fairness checks into routing algorithms also helps, along with regularly evaluating performance and using randomized routing when needed to maintain fairness.

To ensure fair service levels, we need transparent routing criteria so everyone understands how decisions are made. We also have to balance workloads across service agents, avoiding overburdening some while neglecting others. And, importantly, fairness objectives must be built directly into the routing algorithms themselves to prevent discrimination or neglect.

Transparency with customers about AI use is critical. When companies explain how AI is involved, it empowers customers to make informed decisions and builds trust. Hiding AI involvement can have the opposite effect — damaging trust and creating suspicion.

There are several ways to inform users about AI involvement. Clear notifications are a straightforward way to do this. Consent forms give users control before they engage with AI-powered services. User guides can explain how AI features work, and accessible explanations within products ensure customers know exactly what AI is doing and why.

When companies are transparent, it builds trust and improves the user experience. But it's important this is done thoughtfully — we want clear communication that informs, not alarms, users. The right approach strengthens relationships and confidence in the technology.

AI hallucinations happen when AI systems produce false or fabricated information, even though they sound convincing. Misinformation is similar — it's false or misleading content, but it can have serious consequences if users act on it, affecting both individuals and society.

Inaccurate AI outputs raise several ethical concerns. They risk misleading users and causing bad decisions. If errors happen repeatedly, users lose trust in the technology. There's also the question of accountability — who is responsible when AI gets it wrong? And in sectors like healthcare or finance, misinformation can have real, harmful effects.

So how do we guard against these risks? Better model training can reduce hallucinations and misinformation. Verifying AI outputs before they reach users helps ensure accuracy. We also need human oversight and user warnings to manage AI's limitations. And finally, clear regulatory frameworks can guide developers and companies in responsible AI use.

To sum up, ethical AI is about fairness, transparency, accountability, and protecting users. By applying these principles — from personalization and agent routing to customer transparency and misinformation safeguards — we can build AI systems that benefit people and society while minimizing harm.

Chapter 12: Emerging Technologies

Data privacy is a core part of delivering a trustworthy customer experience. Customers expect their data to be handled with care and transparency. Without strong privacy measures, organizations risk losing customer trust, facing regulatory penalties, and damaging their reputation.

The regulatory landscape for data privacy is constantly evolving. Organizations must comply with laws like the GDPR in Europe and the CCPA in California, both of which enforce strict rules on how customer data is collected and handled. They also face the challenge of navigating diverse legal obligations across different countries and sectors. Cross-border data management complicates things further, as data must often comply with multiple regulations simultaneously. It's essential for businesses to stay agile and adapt quickly to new and changing regulations.

To ensure compliance, organizations must weave data protection strategies directly into their customer experience operations. This starts with strong data governance — aligning how customer data is collected, stored, and used within a structured framework. Privacy policies must be integrated into every customer touchpoint, ensuring consent is captured and legal requirements are met. Technology plays a big role here, offering tools that help enforce compliance seamlessly within customer interactions and across CX operations.

Let's start with a quick overview of GDPR and CCPA. GDPR governs data privacy across the European Union, emphasizing strict consent requirements and granting individuals strong rights over their data. The CCPA provides similar protections for California residents, focusing on transparency and giving consumers control over their personal information. Both regulations share common themes — they demand transparency, enforce consent, and promote robust data protection.

A key part of these regulations is customer consent. Consent must be clear, explicit, and fully informed — customers need to know exactly what they're agreeing to. They must also be able to withdraw that consent at any time, easily and without facing any penalties. Organizations need systems in place to manage consent — tracking it, recording it, and respecting customers' preferences.

Technology can help streamline consent management. Automated platforms simplify the process of collecting consent through digital forms, keeping everything updated in real time. These platforms securely store consents, making them easy to retrieve for audits or compliance checks. They also support compliance auditing, ensuring transparency and regulatory adherence. Ultimately, these tools give customers better control over their data and help build trust.

Federated learning is a method where AI models are trained directly on user devices, instead of sending raw data to central servers. This keeps sensitive data on the device, significantly reducing privacy risks. At the same time, it allows AI to provide personalized insights by learning from the data it processes locally — without compromising the user's privacy.

Edge AI takes this concept further by enabling decentralized data processing. Instead of relying on centralized servers, data is processed closer to where it's generated — right on the edge devices. This reduces the risk of privacy breaches, enhances security, and supports compliance with data minimization principles by reducing the amount of data stored centrally.

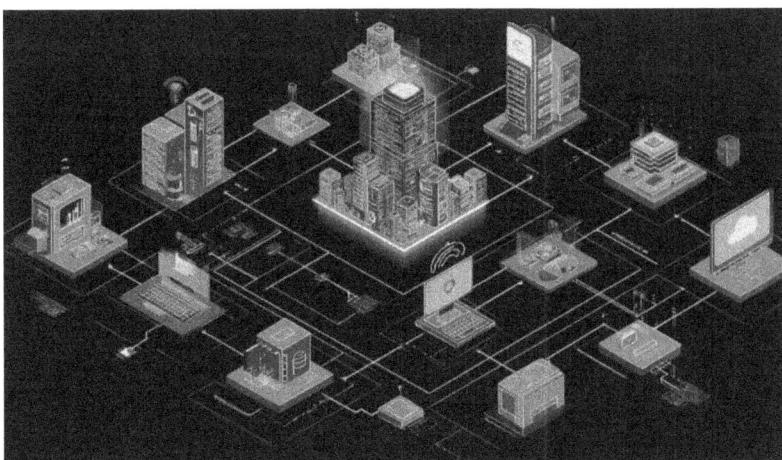

Explainable AI, or XAI, is about making AI systems understandable to humans. In customer experience, this means AI systems that don't operate as black boxes, but rather explain their decisions and actions clearly. This transparency builds trust — both with customers and regulators — by making AI-driven processes easier to understand and scrutinize.

Transparent AI decision-making helps customers feel more secure about how their data is being used. When customers understand why an AI system made a certain decision, they're more likely to trust the system and the organization behind it. This clear communication strengthens customer relationships and builds confidence in AI technologies.

Of course, transparency needs to be balanced with business goals and privacy needs. Organizations must find the right level of openness — enough to build trust, but not so much that it compromises business strategies or security. This balance is key to successful customer experience and compliance management.

Trust engineering is about embedding trust into digital systems by design. This involves applying security principles to protect against threats, ensuring privacy by safeguarding user information, creating intuitive systems through good usability design, and maintaining transparency about how data is used. These principles together help build systems that customers can trust.

To enhance customer confidence, organizations should focus on clear communication — making sure information is straightforward and accessible. They must consistently apply strong privacy practices and implement solid security measures to protect data. Equally important is responsive customer support — addressing questions and concerns quickly to build trust and enhance satisfaction.

Long-term trust isn't something you set and forget. Organizations need to monitor trust metrics regularly, using them to identify areas for improvement. Collecting and analyzing customer feedback helps catch potential issues early and strengthen relationships. And finally, businesses must stay ahead of emerging risks — adapting their strategies to maintain customer confidence over time.

In conclusion, data privacy and compliance aren't just regulatory requirements — they're essential for building lasting customer trust. By integrating compliance strategies into CX, leveraging technologies like federated learning and explainable AI, and focusing on trust engineering, organizations can deliver secure, transparent, and trustworthy customer experiences. This commitment to privacy and transparency is key to long-term success in today's digital landscape.

Chapter 13: Metrics For Measuring AI Impact

Overview of AI's Role in Customer Experience AI plays a central role in enhancing customer experience. It provides personalized support by analyzing customer preferences and tailoring responses in real time. It also automates routine tasks, boosting efficiency and reducing response times. Another major benefit is AI's ability to predict customer needs, allowing for proactive engagement. And most importantly, measuring AI's impact is critical — it helps businesses assess value and continuously improve their customer experience strategies.

Importance of Measurement and Quantifiable Outcomes Measuring the impact of AI is not just about numbers — it justifies investment by providing clear evidence of value. With quantifiable outcomes, businesses can fine-tune their AI strategies for maximum effectiveness. And by tracking improvements, companies can clearly demonstrate

AI's positive effect on customer satisfaction and operational performance.

Defining NPS, CSAT, CES, and CLV Let's define some key metrics. The Net Promoter Score, or NPS, measures customer loyalty based on their willingness to recommend a brand. Customer Satisfaction, or CSAT, reflects how happy customers are with a product or service. The Customer Effort Score, or CES, gauges how easy it is for customers to get support. And finally, Customer Lifetime Value, or CLV, estimates the total revenue a business can expect from a customer over time.

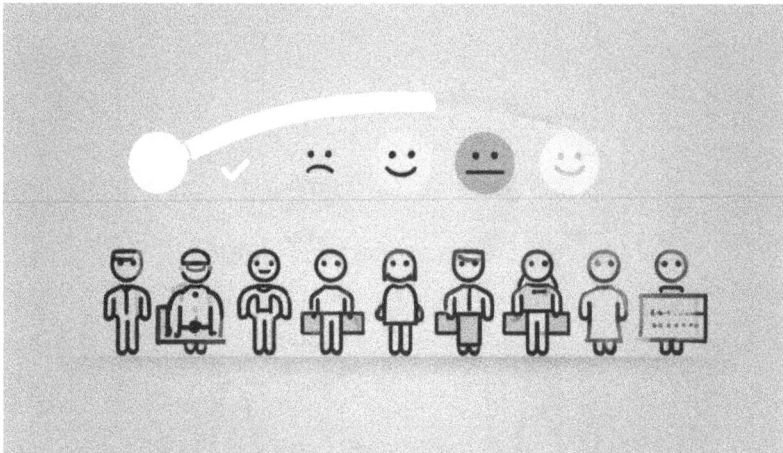

How AI Initiatives Influence These Key Metrics AI directly improves these metrics in several ways. AI chatbots, for example, reduce friction by offering instant, efficient support. Personalized recommendations powered by AI drive customer satisfaction and engagement. And proactive support — anticipating customer needs before they arise — improves effort scores and satisfaction levels. All these initiatives contribute to better NPS, CSAT, CES, and ultimately, higher customer lifetime value.

Case Examples of Improvements Driven by AI Here are some examples of AI-driven improvements. AI significantly reduces customer wait times, boosting service efficiency. It can increase customer satisfaction scores by as much as 20%, reflecting better service quality. And perhaps most importantly, AI-driven enhancements to customer experience lead to higher retention rates, which directly benefit long-term business growth.

Understanding Resolution Time as a Metric Resolution time measures how long it takes to fully resolve a customer issue — it's a key efficiency indicator. AI tools help shorten this time by accelerating problem-solving processes. The faster you resolve a customer's problem, the better their overall experience.

Measuring the Impact of AI on Deflection Rate Deflection rate measures how many customer inquiries are resolved by automated tools without needing a human agent. AI plays a key role here — it increases resolution speed and improves overall support efficiency. By handling common queries, AI frees up human agents for more complex tasks, reducing operational costs while enhancing the customer experience.

Correlation Between AI, Resolution Speed, and Support Efficiency When you deploy AI, you typically see a direct improvement in resolution speed. Faster problem-solving leads to higher customer satisfaction and loyalty. Plus, AI boosts support efficiency, cutting costs and making better use of resources. It's a win-win for both businesses and customers.

Attribution Models for AI-Driven Outcomes Attribution models help businesses understand AI's impact across the customer journey. Multi-touch attribution tracks multiple AI interactions influencing customer decisions. These insights assign value to each AI-driven touchpoint, making it clear how AI contributes to key business outcomes.

Calculating ROI for AI Projects To measure AI's ROI, start with cost savings — look at how much you've reduced operational expenses through AI automation. Then assess revenue growth directly tied to AI-driven improvements. Finally, measure the lift in customer satisfaction that comes from AI-powered personalization and support. These elements together give a clear picture of AI's return on investment.

Challenges and Best Practices in AI Attribution Attributing results to AI can be tricky because other factors often influence outcomes. That's why it's crucial to use robust, high-quality data for attribution models. Also, defining clear KPIs and setting up continuous monitoring helps ensure your attribution is both accurate and meaningful.

Establishing and Updating Benchmarks for AI Performance Benchmarks provide a baseline to measure AI effectiveness. They guide performance goals and show whether AI is meeting expectations. But benchmarks need regular updates to stay relevant as technology evolves. By keeping them current, you ensure they remain a reliable measure of AI success.

Continuous Learning and Model Improvement AI systems aren't static — they learn and improve over time. Continuous learning from new data helps AI stay accurate and effective in changing environments. This ongoing improvement directly translates to better customer experiences and outcomes, as AI becomes smarter and more responsive.

Chapter 13: Metrics For Measuring AI Impact

Industry Standards and Future Outlook Looking ahead, emerging industry standards will help ensure AI systems are transparent, fostering trust with customers. As AI technology advances, we'll see more sophisticated data analysis, deeper customer insights, and even more personalized experiences. The future of AI in customer experience is both exciting and full of potential.

Conclusion To wrap up — measuring AI's impact is crucial for maximizing its value in customer experience. By tracking key metrics, applying solid attribution models, and promoting continuous improvement, businesses can unlock AI's full potential. Remember, AI is not just a tool — it's a strategic asset for enhancing customer satisfaction, driving efficiency, and growing your business.

Chapter 14: Aligning AI With Brand And Experience Strategy

To successfully integrate AI into your brand strategy, it's essential that AI solutions genuinely reflect your brand's identity and values. The goal is not just to adopt AI for efficiency, but to enhance how customers perceive your brand. Every AI interaction should build trust by staying true to your ethical standards and reinforcing your brand's core message.

AI should deliver a seamless and consistent experience across all customer touchpoints. Whether a customer interacts with your chatbot, mobile app, or voice assistant, the brand message and tone should feel unified. This consistent experience strengthens brand identity and ensures customers recognize and trust your brand no matter where they engage with it.

While AI automation boosts efficiency, it's important not to lose the human touch that builds real connections. AI can handle routine tasks, but human interactions bring authenticity and emotional engagement. The ideal approach is to blend both—using AI to scale operations while preserving the human element that defines meaningful brand experiences.

When designing AI to reflect your brand voice, it helps to create AI personas that mirror your brand's tone. This ensures consistency in interactions. Additionally, scripting AI responses to align with your brand voice strengthens customer engagement and loyalty. A well-designed AI personality helps deepen the emotional connection with your audience.

Clear guidelines are critical for AI-generated content. These guidelines ensure that the style, language, and tone remain consistent with your brand identity. They also help AI-generated content comply with legal and ethical standards, protecting your brand's reputation and maintaining customer trust across all platforms.

AI outputs need continuous monitoring to ensure they stay aligned with your brand voice. By regularly reviewing content, you can quickly spot any deviations or inconsistencies. Feedback loops and iterative refinement help improve AI interactions over time, strengthening brand alignment and enhancing customer experience.

Defining clear roles for humans and AI in the customer journey is essential. AI should handle routine and repetitive tasks, improving efficiency, while humans step in for complex situations that require empathy and emotional intelligence. When humans and AI collaborate effectively, they can deliver better problem-solving and higher customer satisfaction.

Training your customer experience teams to work alongside AI is key. Empower your teams with the knowledge and skills to collaborate with

AI tools. This partnership ensures that AI supports—not replaces—human roles, fostering a more effective and adaptable customer experience environment.

AI-generated insights are powerful tools for customer experience professionals. These insights enable data-driven decision-making, allow teams to personalize interactions, and help them proactively address customer needs. By leveraging AI data, CX professionals can deliver more targeted, meaningful, and timely support to customers.

A successful AI strategy starts with a well-defined adoption plan. This includes identifying opportunities where AI can enhance customer experience and ensuring that its integration is purposeful and aligned with business objectives. Strategic planning is crucial for maximizing the impact of AI on customer interactions.

Ethical guidelines and compliance measures must be part of any AI governance framework. This includes protecting customer privacy, ensuring fairness by eliminating bias, maintaining transparency in AI decision-making, and adhering to all relevant legal and regulatory requirements. These practices build trust and protect your organization's reputation.

Continuous evaluation is key to improving AI performance. Regular assessments help identify strengths and weaknesses, and refining AI models over time ensures better outcomes. Iterative improvements enhance both the AI's effectiveness and the overall customer experience, making your AI systems more adaptable and customer-focused.

To become an AI-first organization, you need a culture that embraces innovation. Encourage openness to new technologies and foster a mindset of continuous learning. Using data-driven insights for decision-making empowers your teams to leverage AI effectively and stay competitive in a fast-changing market.

AI integration often requires rethinking existing workflows and operations. Adjusting processes to accommodate AI can improve efficiency and foster better collaboration. This might mean redefining roles to include AI oversight or evolving operational models to fully harness AI's capabilities, ensuring you maximize the value from your AI investments.

Preparing for the future of AI in customer experience means nurturing the right skills in your teams. Focus on building AI literacy, developing strong data analysis abilities, and encouraging the use of customer-centric technologies. This ensures your staff are equipped to work effectively with AI and deliver enhanced customer service.

In conclusion, aligning AI with your brand and experience strategy isn't just about technology—it's about ensuring every AI interaction supports your brand identity, enhances customer trust, and delivers meaningful experiences. By balancing automation with human empathy, setting clear guidelines, and fostering a culture of continuous learning, you can build AI-first customer experience organizations that drive real value.

Chapter 15: Emerging Tech In Cx

The landscape of customer experience is changing fast, driven by technological advancements. Artificial intelligence helps businesses personalize and anticipate customer needs. Machine learning enables systems to learn from customer behavior and continuously optimize services. Natural Language Processing allows companies to better understand and respond to human language across various channels. And automation helps streamline repetitive tasks, making customer service more efficient and consistent.

Several key factors are driving innovation in customer experience. First, growing customer expectations — today's customers want seamless, personalized interactions whether online or in person. Second, the availability of vast amounts of data allows businesses to gain deeper insights into customer behavior. Third, AI capabilities are advancing rapidly, enabling smarter automation and personalization. And finally, competitive pressures push companies to innovate quickly to stay ahead in the market.

Emerging technologies offer great opportunities to enhance customer experience — whether through personalization or efficiency. However, they also come with challenges. Data privacy remains a major concern as businesses handle increasing amounts of customer data. Integration complexities arise when new tech must work with existing systems. And workforce impact is significant, as roles evolve and employees need to adapt to new tools and workflows.

Self-learning customer experience platforms are systems that can adapt and improve over time without human input. They analyze customer behavior patterns and optimize processes automatically, boosting both efficiency and effectiveness. Most importantly, they personalize experiences uniquely for each customer, increasing engagement and satisfaction without manual adjustments.

Machine learning and AI play a central role in enabling adaptive customer experiences. These systems process massive amounts of data and predict customer needs in real time. They also allow businesses to adjust customer responses dynamically as interactions unfold, creating a more personalized and responsive experience.

There are many practical ways businesses are using self-learning platforms. For example, interactive chatbots that improve through ongoing interactions can enhance support and engagement. Personalized recommendations help increase sales by aligning offers with customer preferences. Automated customer journeys allow businesses to optimize marketing and engagement strategies in real time. But success depends on continuous training and seamless integration with existing systems.

Understanding customer emotions is becoming a key part of customer experience. Technologies like voice tone analysis help detect emotional cues in speech. Facial expression recognition allows systems to interpret real-time emotional states. Text sentiment analysis helps identify emotions expressed in writing. Together, these tools help businesses engage with customers in a more meaningful and satisfying way.

Empathy-driven AI brings a human-like touch to customer interactions. By fostering more natural, emotionally aware engagements, AI can help build trust with customers. It also reduces frustration by responding appropriately to emotional cues. Ultimately, empathy AI can lead to stronger customer relationships and increased loyalty.

We're already seeing emotion-aware systems in action. In call centers, emotion detection helps prioritize and route calls based on a customer's mood. Virtual assistants can adapt their tone and responses dynamically, improving the quality of interactions. And in retail, support bots that recognize emotions provide more sensitive and effective assistance.

AI avatars are virtual, human-like figures controlled by AI, creating immersive experiences. Digital twins go a step further — they're highly accurate, data-driven models of real people, used for simulation and personalized interactions. Both technologies are changing how businesses engage with customers.

These technologies have several applications in customer experience. Virtual assistants powered by AI provide real-time help and automate customer responses. Personalized marketing campaigns can be tailored more precisely using AI insights. Training simulations offer realistic learning environments for employees. And with AI-driven systems, businesses can offer 24/7 customer support that closely mimics human interaction.

AI avatars and digital twins offer many benefits, from increased personalization to operational efficiency. However, they also raise ethical questions, such as how data is used, consent, and maintaining transparency with customers. Businesses must navigate these considerations carefully to ensure trust and fairness.

Agent assist technologies are designed to support human customer service agents. AI-powered tools offer real-time suggestions, improving response times and accuracy. Automated knowledge retrieval systems help agents access information quickly. And real-time analytics provide insights during interactions, enabling agents to deliver better service with empathy and understanding.

We're now seeing a transition from agent assist toward more automation. As AI capabilities advance, autonomous customer interactions—like chatbots and virtual agents—handle more routine queries on their own. This reduces the need for human intervention, especially for repetitive tasks, and improves efficiency.

While automation brings benefits like increased efficiency and service availability, it also impacts the workforce. Employees will need reskilling to manage and work alongside these new technologies. At the same time, businesses must ensure automation doesn't compromise customer trust or service quality, which are essential to long-term success.

To conclude, emerging technologies are revolutionizing customer experience, offering both exciting opportunities and important challenges. Businesses that adopt these innovations thoughtfully can deliver more personalized, efficient, and emotionally aware service. But success depends on balancing technological advancements with ethical considerations, data privacy, seamless integration, and a skilled, adaptable workforce. Thank you for joining me in this session on emerging technologies in CX.

Chapter 16: Immersive And Multimodal Cx

Let's start by defining immersive and multimodal customer experiences. Immersive CX uses technologies like virtual reality, augmented reality, and spatial interfaces to create rich, interactive environments where customers can engage on a deeper level. Multimodal CX, on the other hand, allows customers to interact using a combination of voice, text, and visual inputs—making their experience feel more seamless, natural, and intuitive.

Why does this matter for today's customer journeys? First, immersive experiences allow brands to tailor interactions to each customer's preferences, making them feel more valued. Second, by engaging customers through multiple sensory channels—like sight, sound, and even touch—you enhance overall engagement. And finally, these richer experiences can boost customer satisfaction, build loyalty, and drive conversions.

Of course, adopting immersive CX isn't without its challenges. The technology itself can be complex, requiring thoughtful design and integration. Connecting these systems with your existing infrastructure can be tricky, but it's crucial for a smooth customer experience. And privacy is a major concern—businesses must address this proactively. But the upside is huge: overcoming these hurdles can open up powerful new ways to engage customers and stand out from competitors.

AR and VR are reshaping the shopping experience. Immersive environments let customers explore products in interactive, lifelike ways. AI takes this further by offering personalized, real-time recommendations based on customer behavior and preferences. Plus, AR enhances visualization by giving customers detailed, contextual information about products—helping them make better buying decisions.

AI also powers personalization and predictive analytics in immersive retail. By analyzing customer data, businesses can predict shopping habits and tailor experiences accordingly. Immersive environments can adapt dynamically to customer preferences, creating a personalized experience every time. This boosts engagement, making the shopping journey feel more relevant and compelling.

Multimodal interaction is all about giving users flexible ways to engage. Voice commands provide a hands-free, natural interface. Text input remains important for precision and control. Gesture recognition adds an intuitive layer of interaction, especially in AR/VR contexts. And visual cues—like icons or animations—help guide users and provide feedback. The result is a more engaging and user-friendly experience.

A key benefit of multimodal interfaces is seamless switching between modes. Users can transition smoothly from voice to text to gestures, all within the same experience, without losing flow. By combining different input methods, businesses can create more intuitive and efficient interactions across a variety of platforms and settings.

We're seeing growing adoption of multimodal interfaces across industries like retail, healthcare, and customer service. Consumers are increasingly expecting natural, conversational, and context-aware interactions. This trend is pushing companies to evolve their interface designs to meet these rising expectations.

Spatial computing brings together digital information and the physical world. By blending these elements, it enables highly immersive experiences. Using sensors and AR/VR technology, spatial computing systems can perceive and interact with the real-world environment, creating a richer, more dynamic user experience.

Contextual AI plays a big role in understanding user intent. By analyzing environmental data—like location or surroundings—AI can offer more relevant responses. It also interprets behavioral data to personalize interactions in real time. This allows businesses to provide proactive, tailored responses that feel natural and highly relevant to each customer.

When combined with IoT and smart environments, spatial computing becomes even more powerful. Devices equipped with spatial computing can interact with their surroundings intelligently. Contextual AI adjusts device behavior based on both the environment and user preferences. Connected IoT devices work together within smart environments to deliver seamless, integrated experiences for users.

Digital experience twins are virtual replicas of customer interactions and environments. They allow businesses to simulate and analyze how customers will engage with products or services. This technology helps refine processes and strategies by providing valuable insights in a risk-free, virtual setting.

Simulated CX labs take this a step further by using digital twins to test new features and customer journeys before they go live. This reduces risks, helps identify potential issues early on, and improves the overall effectiveness of customer experience design—ensuring better outcomes when solutions are launched in the real world.

Ultimately, digital twins and simulated environments offer a powerful tool for continuous improvement. By testing, refining, and optimizing CX strategies in a controlled setting, businesses can enhance user satisfaction and stay ahead in the competitive customer experience landscape.

To wrap up, immersive and multimodal CX is a game-changer for businesses looking to engage customers in new and meaningful ways. From AI-powered AR/VR experiences to multimodal interfaces, spatial computing, and digital twins, these technologies offer incredible opportunities. By embracing them thoughtfully, businesses can create richer, more personalized customer journeys that drive loyalty and growth.

www.ingramcontent.com/pod-product-compliance
Lightning Source LLC
Chambersburg PA
CBHW060627210326
41520CB00010B/1499